THE F
INDUS

Peter Walter

SHIRE PUBLICATIONS

First published in Great Britain in 2010 by Shire Publications Ltd, Midland House, West Way, Botley, Oxford OX2 0PH, United Kingdom.
44-02 23rd Street, Suite 219, Long Island City, NY 11101, USA.

E-mail: shire@shirebooks.co.uk   www.shirebooks.co.uk

© 2010 Peter Walter

All rights reserved. Apart from any fair dealing for the purpose of private study, research, criticism or review, as permitted under the Copyright, Designs and Patents Act, 1988, no part of this publication may be reproduced, stored in a retrieval system, or transmitted in any form or by any means, electronic, electrical, chemical, mechanical, optical, photocopying, recording or otherwise, without the prior written permission of the copyright owner. Enquiries should be addressed to the Publishers.

Every attempt has been made by the Publishers to secure the appropriate permissions for materials reproduced in this book. If there has been any oversight we will be happy to rectify the situation and a written submission should be made to the Publishers.

A CIP catalogue record for this book is available from the British Library.

Shire Library no. 576   •   ISBN-13: 978 074780 753 7

Peter Walter has asserted his right under the Copyright, Designs and Patents Act, 1988, to be identified as the author of this book.

Designed by Ken Vail Graphic Design, Cambridge, UK and typeset in Perpetua and Gill Sans.
Printed in China through Worldprint Ltd.

10 11 12 13 14   10 9 8 7 6 5 4 3 2 1

COVER IMAGE
Making Valenki boots, c. 1940.

TITLE PAGE IMAGE
A heavy duty flat hardener used for making thick felt at Bury and Masco Industries Limited, Baltic Mill.

CONTENTS PAGE IMAGE
Mini pods in felt by Deborah Roberts.

ACKNOWLEDGEMENTS
It has taken quite some time to prepare this book and many people within the felt industry have helped me on the journey, particularly: Mrs. Ashworth, Mr. Barker, H. Clark, N. Clegg, Mr. Crease, T. Crook, W. Davis, C. Dyson, K. Entwistle, R. Entwistle, D. Fearnley, P. Fuller, F. Haig, D. Hamer, W. Law, J. J. Livesy, Mrs. Lord, J. Lovell, B. Neary, F. W. Nicholls, K. Ritson, J. Roberts, D. Rodgers, J. Rostron, E. Sagar, T. Simpson, S. Smith, W. Suart and E. Townsend. The book would not have been possible without the encouragement, support, and archival material that past directors of the major British felt companies have given me, especially: A. Coupe, A. F. Ferguson, P. S. James, L. Notley, and R. J. Ponting. I am indebted to M. Samuelson of Garnett Bywater Limited, who allowed access to their daybooks and donated their photographic archive of felt making machines when they ceased their manufacture. Thanks too to the Feltmakers Association for information about today's felting scene. It is mainly through the supporting research work by Ann Baxendale and the moral support of my wife Liz that I have at last completed this book.

PHOTOGRAPH ACKNOWLEDGEMENTS
T. G. Ames, page 17; Mrs. M. P. Ashworth, page 20; Mary E. Burkett OBE and Dr Jarman, page 6; R. Giddins, page 30; Mrs Hamer, page 37; Helmshore Mills Textile Museum, page 36; Ewa Kuniczak, page 4; Lancashire County Library, SE Division, Rawtenstall, pages 19, 21, and 22; Deborah Roberts, contents page; Sheila Smith, page 8; Wandle Industrial Museum, page 25; and The Worshipful Company of Feltmakers of London, page 6.

Shire Publications is supporting the Woodland Trust, the UK's leading woodland conservation charity, by funding the dedication of trees.

CONTENTS

INTRODUCTION	5
WHAT IS FELT?	9
INDUSTRY HISTORY	13
PROCESSES	29
PRODUCTS	41
LIFE IN THE FELT MILLS	51
FURTHER READING	54
PLACES TO VISIT	55
INDEX	56

INTRODUCTION

Felt is an amazing material: its history, properties and uses are unparalleled amongst textile materials and even today it could replace many of the synthetic materials in common usage. Although its origins are lost in prehistory there is a clear record of how it was made and what it was used for right back to Roman times and there is no doubt it had a special place in most cultures, particularly in the East. It is well known that the Mongols revered felt and attributed magical powers to it, as witnessed by the totem felt dolls that they hung outside their tents. But the Mongols were also very practical and used felt to supply all the comforts of a nomadic life, covering their tents, or yurts, with felt and their floors with felt carpets. They obviously knew first hand the key properties of felt: it is light in weight, soft and warm to the touch, durable, water resistant and with great heat insulation properties – all ideal attributes for nomads travelling in harsh environments. Recent evidence from the frozen tombs of Siberia in Pazyryk, Russia, has shown that the Mongols were by no means unique in their use of felt, since excavations there have revealed felt artefacts of stunning design and ornamentation that have been used for a whole range of the domestic needs of the time, such as hats, apparel, toys, saddle blankets and harnesses. The manufacture and use of felt is now known to have been widespread throughout Asia and Russia and even today there are areas where the old methods of felt making are still being practised. It is the Eastern countries that have maintained the ancient and traditional art of felt making, and it is still flourishing, long after the felt industry in the UK disappeared in 1993. It is thanks to pioneers like Michael and Veronika Gervers and Mary E. Burkett, who travelled extensively through Turkey, Anatolia and Iran searching out felt makers there, that the extent and skills of these Eastern felt makers has come to light. Their travels through Asia have shown the rich diversity of felt making throughout the region and their publications and presentations have had a major impact on today's textile designers and artists who are now part of a new and growing artistic movement that is exploring the infinite design potential of hand crafting felt.

Opposite:
Rainbow butterfly felt by Ewa Kuniczak.

THE FELT INDUSTRY

Traditional felt making in Turkestan.

All this rich variety of uses and the ability to manufacture seems to have been lost to the Western countries right up to the nineteenth century, for the only felt manufacture of any consequence was hat making, which eventually built up into a sophisticated and widely distributed industry. However, hat making is only loosely allied to the woollen felt industry since hatters preferred to use fur fibres, like beaver and coney (rabbit), in preference to wool because these gave a finer finish to the hat. An indication of its importance was the formation of the Worshipful Company of Feltmakers of London, a guild to which the hat makers had to belong.

Coat of arms for the Worshipful Company of Feltmakers.

Like most craft organisations the hat felt makers adopted a patron saint, choosing St Clement to intercede for them. For good measure they featured him in an apocryphal story about the origin of felt, in complete denial of the prehistoric origins for felting. Although there are many variations of this story the most popular one tells of how St Clement was on a pilgrimage and his feet became sore after he had travelled some distance. He spotted some loose wool caught on nearby bushes and stuffed the fibres into his sandals to cushion them. On reaching his destination the heat, sweat, and movement of his feet had turned the wool into felt. Even though the truth is of course far removed from this in both time and substance, the symbolic image is so strong it persisted right up to modern times and one of the largest felt makers in Britain, Bury Felt Manufacturing Company, had a picture of St Clement etched into the window of its boardroom in 1956.

It was the mechanisation of felt making that was the great innovation of the West. It appeared to come from nowhere in 1840 complete with its own technology and terminology and with a scientific approach that was typical of the Industrial Revolution, far removed from the free flowing approach of Asia. In fact the inventor of the felt industry, T. R. Williams, was connected with F. McNeill and Co. supplying asphalt felts for roofing, for which there was significant commercial advantage in being able to produce continuous lengths. He used this experience to design and build a process for manufacturing woollen felt. At this point the felt industry diverged away from the asphalt-felt industry, though the legacy of their joint creation remains, since after this time asphalt felt became universally known as 'roofing felt'.

When the new continuous felt making process and its products were revealed it was hailed as a textile revolution, finding an instant use as carpeting. Through a strong patent, manufacturing was restricted to just two companies in Leeds for the period from 1840 to 1854. It was only when the patent expired that there was a proliferation of companies throughout the country as entrepreneurs tried to realise the potential profits made by the market leaders. For a while Leeds remained the centre for the expansion of the industry, but gradually companies set up in the Rossendale Valley and flourished at the expense of the Leeds industry, superseding it entirely in 1904. This was largely due to the development of the slipper industry in Rossendale that not only boosted felt sales, but also made the area a centre for the British shoe industry. This was not the only innovation made by the felt makers of Rossendale: they introduced needle punching as an alternative to felting, which became another completely new non-woven industry for the synthetic fibres that became readily available in the 1950s.

Glass etching of the window in the boardroom of Bury Felt Company, depicting St Clement's story.

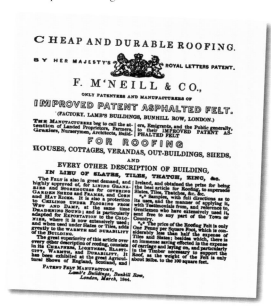

Advertisement for McNeill's roofing felt in 1844.

Leafy wrap in felt by Sheila Smith.

Companies in other parts of the country competed by specialising in niche markets such as piano felts or dense felts for polishing, whilst some produced coarse insulating felt from hair products that were more closely associated with the asphalt industry than woollen felt. However, it was inevitable that the felt industry had to rationalise as it came under pressure from other materials. From 1904 onwards there were continual mergers and mill closures culminating with the formation of Bury and Masco Industries in 1962 and Bury Cooper Whitehead in 1980, which in the ultimate rationalisation closed altogether in 1993.

Ironically, as the industrial felt industry declined the hand felting craft movement has grown from strength to strength and the Feltmakers' Association, which was formed in 1978, now has over eight hundred members worldwide committed to furthering the development of handmade felt. It is possible to see in this enthusiastic revival all the vibrancy that was once common in a now lost commercial industry.

Today there is nothing left of the woollen felt industry in Britain. All the mills, except for E. V. Naish, have been demolished or turned into housing estates; the machinery has gone or been exported, together with all its technology and skills. What follows in this book is an attempt to document and preserve a little of the achievements, technology, and products of this once vibrant industry, and to record the personalities that shaped it. If felting had been invented today it would be proclaimed a new wonder material; an eco-friendly sustainable product that has great potential for substituting oil-based products. Maybe one day we may see its return.

WHAT IS FELT?

THERE ARE MANY different descriptions of felt such as hair felt, roofing felt, woven felt, hat felt, woollen felt, and needlefelt; and there is even a carbon felt used in the heat shield of the American Space Shuttle.

Hair felt is made from assorted animal fibres, usually cow or horse hair, made into a cheap, coarse self-supporting non-woven material that was suitable for impregnating with asphalt to make a waterproof material that came to be known as roofing felt. Woven woollen felt, also known as baize, is a fabric woven from woollen yarn that is felted in a fulling machine to tighten it and give it the smooth matt surface that is characteristic of a felt. However, it was the hat industry that set the standard for the finest surface appearance of a felt by using ultra fine hairs of rabbits and beavers, which were notoriously difficult to felt without the use of toxic chemicals. The manufacture of hats was also the first commercial non-woven industry and being a hand-crafted batch operation rather than a continuous process was organised by the Worshipful Company of Feltmakers of London, which established its traditions and maintains them to the present day.

When the continuous process for making woollen felt was invented the resultant non-woven material was known as 'Patent Woollen Cloth' to distinguish it from baize, because it dispensed with the need for spinning, weaving, or knitting. It was only later that it was referred to as 'felt', eventually becoming a household name synonymous with all forms of non-woven fabrics with a matt surface.

Makers of woollen felt also made hair felts that, compared to wool, they found difficult to process and they supplemented the felting by punching the hair with barbed needles using a special needle-punching machine to entangle the fibres. As synthetic fibres became available, they realised that these would not felt like wool but they could be made to look and perform like a woollen felt if they needle-punched them as though they were hair felts, thereby creating the modern needling process for producing non-woven materials.

However, all these forms of felt are very different materials and the only common factor between them is the fact that their fibres are more or less

A woollen felt magnified five hundred times, showing the characteristic tangling of fibres.

randomly tangled together, with only animal fibres having the natural ability to do this on their own.

Animal fibres can felt together because their surface is made up of scales. These scales make them smooth one way and rough the other and give them different frictional properties when rubbed in opposite directions, an effect known as the 'directional frictional effect' or 'DEF' for short.

Take any hair, animal or human, and rub it between finger and thumb and it will 'walk' in one direction only: turn the hair around and it will go in the opposite direction. It is the same for a bundle of wool: each fibre will travel in a different direction and tangle with the other fibres to make a small dense felt ball, and this was how the wool buyers of the felt industry used to assess the felting qualities of the wools they were about to buy. Interestingly, if the fibres were all aligned in the same direction they would all travel in the

Electron microscope image of a single wool fibre magnified 2,250 times, showing the surface scales.

same direction and therefore would not tangle – which is why sheep are not bundles of felt on legs.

Other factors are also important in making a felt, particularly the curliness or 'crimp' of the fibres, which makes them act like corkscrews when they are vibrated. During felting they twist around each other and amplify the tangling, the fibres having the most curl giving the tightest felt.

Heat, moisture and pressure are also vital for making a felt, since without these conditions very little felting takes place. When washing clothes there is plenty of heat and water, and since they are constantly rubbed together during washing they are under the perfect conditions to give the right vibration and pressure for felting. On the other hand, when wearing woollen clothes there is never enough heat, moisture, pressure or vibration to allow the fibres to felt together.

Heat and moisture swells wool fibres, raises their surface scales, makes them more supple, and lubricates them so that they can move easily and lock together into a compact mass. However, felting cannot take place unless the fibres are brought together so that they can interact in the first place and this is done by applying pressure of exactly the right amount: too little and the fibres are not close enough; too much and the fibres are so tightly packed that they cannot move. It is the control of all these factors in a continuous process that was the fundamental achievement of the early felt makers, and it was this that led to the formation of the felt industry.

Wool fibres from a fleece showing the curliness or 'crimp'.

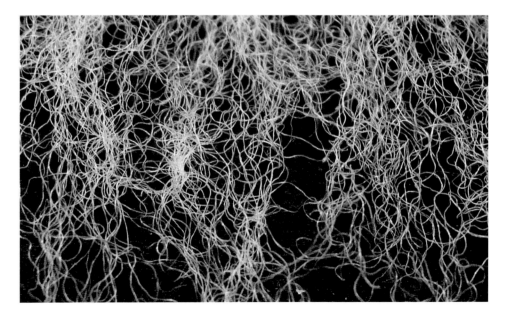

INDUSTRY HISTORY

ORIGINS

The felt industry owes its origin to a marine mollusc in the Caribbean known as the Teredo worm, which had an appetite for boring into the keels of wooden sailing ships. This was a serious problem for the eighteenth-century British Navy, when ships that returned from a tour of duty in the tropics were found to be falling apart. The problem was resolved by plating a ship's hull with copper plate, called 'copper bottoming', but the technique was only partially successful and it needed a layer of asphalt-impregnated woollen or hair felt between the copper plate and the wooden hull to exact a full solution. This created a vast new market for asphalt-impregnated felt, which outstripped the supply because the felts had to be made in relatively small sheets by traditional hand-made hat-felting techniques. Some early entrepreneurs, like William Abbott, invented mechanised ways of felting in 1829 to replace the drudgery of working the fibres by hand, but it was still very much a batch process.

At the same time others were appreciating the technological properties of impregnated felt for engineering uses. Foremost of these was William Borradaile, whose felts were used in many significant civil engineering applications such as in the laying of railway sleepers. Brunel even used it in the construction of the Menai Bridge.

It was in 1839 that the first breakthrough in continuous manufacture occurred through the partnership between the company of F. McNeill and a professional inventor known as Thomas Robinson Williams, which gave rise to a patented process for producing asphalt-impregnated felt using animal and vegetable fibres.

Williams licensed the process to Forbes McNeill, who was quick to realise the commercial advantage of manufacturing long lengths of material, since it opened up a new market for him in roofing felt as an alternative to the thatch in common use at the time. In the meantime, Williams went on to develop his process in a totally new direction with a new partner and between them they founded the woollen felt as the first mass-produced non-woven material.

Opposite: Cooper and Company displaying felt bicycle handles at the Birmingham Cycle Show, 1903. Cooper and Co. was a small specialist felt maker that started by manufacturing felt handles for bicycles by cutting thick dense felts and turning them cylindrically on a lathe according to a patent that it licensed from H. A. Ollerant. The company later fabricated its own unique range of thick, dense felts into a wide variety of different forms.

THE FELT INDUSTRY

McNeill's promotion of roofing felt in 1846.

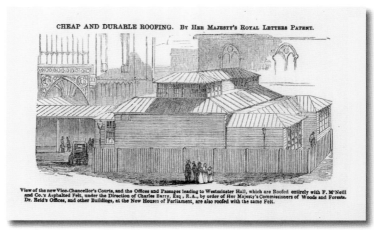

Abbott's patent for hardening felt by moving rollers over fibres on a table.

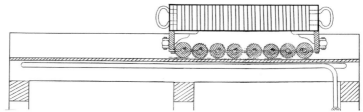

Williams's patent batt frame for winding a batt of wool fibres. The red denotes the supporting conveyors.

The impetus and funding for Williams's new process came from a wealthy wool merchant in Leeds called John Wilkinson, who had seen the potential of making a felted woollen fabric without spinning yarn or weaving cloth. This new partnership led to the patent in 1840, numbered GB 8387, that defined the felt industry and its processes for the next 150 years. It disclosed a machine for making long lengths of carded wool, known as a batt frame, and described the method of felting the fibres together in a process he described as 'hardening'.

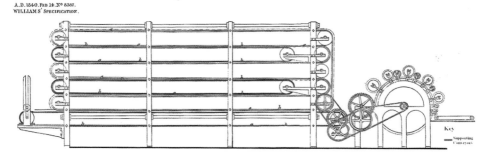

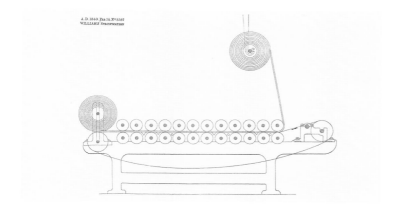

Williams's patent roller hardener for the first felting stage.

The process proved highly successful for manufacturing felt carpet and Wilkinson and Williams set up The Patent Woollen Cloth Company, with production at Elmwood Mill in Leeds and Borough Road in London, to exploit the invention. They ran the company for four years, building it up to sales of £60,246 before selling out to a consortium of businessmen. As a result of his association with Williams, John Wilkinson was granted a licence by the new company owners to continue manufacturing felt on his own account at St Helens Mill, also in Leeds.

T. R. Williams in the meantime went back to join F. McNeill, to continue developing the roofing felt market; the company flourished long after Forbes McNeill's death in 1845.

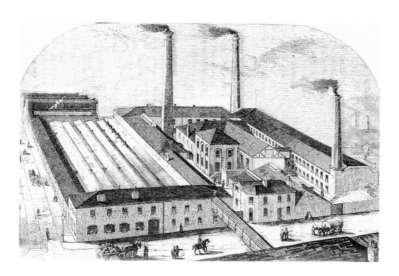

John Wilkinson's mill at St Helens in 1861.

THE LEEDS FELT INDUSTRY

For the fourteen years after the patent was filed until 1854, John Wilkinson and The Patent Woollen Cloth Company had a monopoly over the continuous production of woollen felt, particularly for supplying wide-width printed carpets, receiving recognition at the Great Exhibition of 1851 as gold medal winners. The Patent Woollen Cloth Company concentrated on carpets and created a marketing operation called Royal Victoria Carpets, whilst John Wilkinson pioneered other uses such as saddle felts (numnahs) and a dense felt for gun wads. Both companies had adequate felting capacity but were sometimes short of printing facilities, which they then outsourced by sending their white felt to the Rossendale Valley to be printed.

When Williams's patent expired in 1854 a host of new companies grew up all over England, although the majority were still centred in Leeds. The most notable of these was Carr and Butterworth at Highfield Mill on Dewsbury Road, who until 1878 made the same products as John Wilkinson, in direct competition.

This long-standing partnership eventually split up and George Butterworth continued to make both woollen and hair felt until around 1890 whilst George Carr opted to merchant felt rather than manufacture it; this he did until 1900. Making and selling felt seems to have been a perilous business, because many companies set up facilities only to disappear a few years later; in Leeds, ten companies were established in the space of forty years, each lasting no longer than five years.

Highfield Mill in 1980.

The development of the felt industry in Leeds was aided by virtue of it also being a centre for the production of textile machinery, particularly flax and woollen processing equipment. Significantly, the feltmakers set up their businesses in the same area of Holbeck as the machinery manufacturers.

The first felt machines were undoubtedly made by Taylor Wordsworth, the leading foundry and engineering company of that era, and it was not until 1895 that the company was challenged by William Bywater, who began to specialise in felt making machines. It was Bywater, through one of his engineers, Beanland, who went on to produce the Bywater-Beanland flat hardener. This machine became the workhorse of felt making, and dominated the industry. There was always a close association between Bywater and the later Rossendale felt makers and, through their association with the Mitchell Brothers, they built the first needle-punching machine in the United Kingdom for making needlefelt, thereby initiating and commanding a totally new industry.

Carr and Butterworth advertisement, 1861.

THE ROSSENDALE FELT MAKERS

Edward Rostron is credited as the first to manufacture felt in Rossendale in 1854. Being a merchant, he initially bought in white carpet felt from Leeds, had it printed locally, and sold it on as his own product, eventually becoming prosperous enough to build his own production process at Myrtle Grove Mills, Waterfoot in Rossendale. As he prospered he became a pillar of the local community and a benefactor of St Nicholas Church at Newchurch, which became known as the felt makers' church because so many felt makers worshipped there. After his death, Roland Rawlinson took over the mill and manufactured hair felt there.

By 1860 the felt industry was firmly established in Rossendale with a significant investment coming from James Barcroft, who built sizeable mills at Siss Clough; his own printing works were at Todd Carr, situated on the adjacent site. He too established his business on felt carpet production, which was still a rapidly expanding market due to the significant advantages over woven carpet, being almost three times as wide and half of the cost. James Barcroft remained a dominant force in felt carpet production right up to 1897 when he sold his business to the Mitchell brothers.

Edward Rostron, the first felt manufacturer in Rossendale.

THE FELT INDUSTRY

Position of the felt mills of Rossendale.

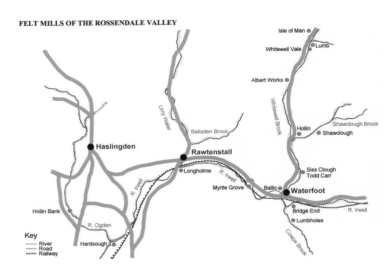

Below: Colonel Thomas Mitchell, one of the three brothers who dominated the felt industry at the turn of the nineteenth century.

Below right: James Barcroft's mill at Siss Clough with Todd Carr Print works.

The Mitchell brothers, Thomas, Robert John Chadwick, and William, were the most entrepreneurial of all the felt makers: they were large in every way, physically, commercially and intellectually. All of them were athletes of Olympic standard and amateur champions of England; Thomas in particular was known to have been of prodigious strength, rising to the rank of colonel as a volunteer in the East Lancashire Regiment.

In 1860 their father, John, first set up business in Waterbarn Mill in Stacksteads, near Waterfoot, as John Mitchell and Sons. In 1865 he expanded

INDUSTRY HISTORY

Mitchell Brothers letterhead showing Albert Works in the 1890s.

Front view of Broadhead's needle loom pioneered in Britain by the Mitchell Brothers.

into Albert Works on the River Whitewell, buying the mill at auction for £3,810 when the Rossendale Printing and Dyeing Company failed. By 1876 the brothers were trading in their own right as Mitchell Brothers and when the Limited Companies Act came into force they became Mitchell Brothers (Waterfoot) Limited in 1893. The brothers were as clever at marketing as they were at manufacturing and soon had offices and warehouses in Leeds, London and Manchester.

If they discovered a new technique they invested in it; where there was a market opportunity but no process, they invented one; and through a series of patents they succeeded in doubling the width of carpets to 144 inches wide, introduced needle punching for coarse felt, and even invested in looms to make woven tapestry carpets.

Significantly, they engaged Henry Rothwell to work for them at Baltic Mill, pioneering the production of felt slippers, but they later sold the business to James Holt when they started losing felt business to other slipper producers such as Sir Henry Trickett, who expanded faster than they did.

By 1870 there was a scramble by the woollen manufacturers to set up felt carpet production, encouraged by the felt machinery manufacturers, though in many cases they attempted

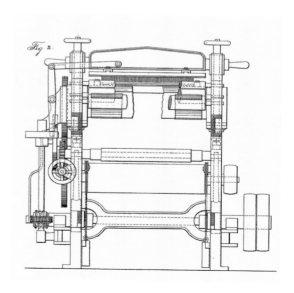

19

THE FELT INDUSTRY

to use waste fibres and broken up rags to maximise profits without the knowledge and expertise to make a good felt. They also lacked their own printing and marketing facilities and were so small they did not have the economies of scale to compete with James Barcroft, Edward Rostron and the Mitchell Brothers. So companies like Captain G. W. Law at Baltic Mill, Ingham Taylor, John Tattersall, Joseph Stanfield, John William Rothwell, and John Wadsworth of Plunge Mill, all went bankrupt within four years after setting up. This only served to make the other established felt makers stronger, since they were able to buy the nearly new machinery at low prices at the liquidation auctions. However, two small companies did manage to survive and challenge the large felt producers.

The first was Richard Ashworth, the son of local celebrity Deborah Ashworth, who started production very modestly by renting Shawclough Mill in 1869 with financial help from his mother. Despite a disastrous flood he managed to salvage his business and expand into the larger Bridge End Mill.

He seems to have had an equal ability for marketing as well as production and sold his felts both under his own name and also through a marketing company in Manchester called A. V. Humphries, which he eventually owned. So successful was he that, in 1896, he was able to buy Longholme Mill in Rawtenstall from William Sutcliffe, the corn millers. Being an astute businessman he also bought the land alongside the mill and allowed the Lancashire and Yorkshire Railway to build sidings on it, so that he had the benefit of shipping his felts by rail directly from his factory to anywhere in the country.

The other successful felt maker was William Stansfield, who started making felt on a small scale at Lumb Holes Mill on the River Irwell, specialising in the manufacture of coarser felts for industrial uses. He was a quiet, determined man and when faced with ruin after his mill at Lumb Holes was almost destroyed by fire, he moved his business into Baltic Mill in 1893 and carried on expanding whilst rebuilding the burnt out mill.

Apart from being a Justice of the Peace he was also endowed with the entrepreneurial spirit, embracing any new technology that could provide advantage, and as such he was one of the first mill owners to install electricity in his mill. He was so enthusiastic about it that he had a line taken from the mill generator to his house in order to light up his home.

The time from 1880 to 1900 was one of considerable turmoil for the felt industry with the impact of Erasmus Brigham Bigelow's carpet power loom beginning to erode the felt carpet market, as woven carpet prices

Richard Ashworth, a leading feltmaker of the Rossendale Valley and co-founder of the MASCO felt combine.

INDUSTRY HISTORY

started to fall and the widths increased to a level that challenged felt's market advantage. Through all this the Mitchell Brothers were prospering because they had such a wide range of products and they clearly saw an opportunity to expand in a weakened market through a series of takeovers of mills and businesses. With the takeover of James Barcroft in 1897 they doubled the size of their business, owning virtually every mill on the Whitewell river.

By 1904 the Leeds felt industry had collapsed and The Patent Woollen Cloth Company was the last remaining felt maker in Leeds, but when they too went into liquidation that year, felt making in Leeds ceased. With just a handful of felt companies left in the United Kingdom, it must have occurred to the Mitchell Brothers that it was in their grasp to form a monopoly. In what looks to be a carefully organised plan, the Mitchells approached their major competitors with a view to combining together into a new company to be known as Mitchells, Ashworth, Stansfield and Company Limited. Having established an agreement of intent with Richard Ashworth and William Stansfield, the three companies set about taking over as many felt companies as they could. Thomas Mitchell bought The Patent Woollen Cloth Company from the receiver, and Richard Ashworth and William Stansfield took over the Hardsough Felt Manufacturing Company, while Thomas Mitchell bought out the Boulinikon Felt Company Limited, a little-known company in Scotland

Baltic Mill in 1950.

THE FELT INDUSTRY

Letterhead of Mitchell Brothers after the merger showing the medals won by The Patent Woollen Cloth Company.

making floor coverings, and then merged them into the new company. Included in the merger were three textile-clothing companies: two of them were merchants, Birtill & Blaikie and William Agar, and one of them was a manufacturing unit trading as The Featherweight Pad Company.

It was on 28 October 1904 that Mitchells, Ashworth, Stansfield and Company Limited was created as the largest felt company ever, having a net worth of some £650,000 and controlling over 80 per cent of the United Kingdom felt industry: almost a monopoly, but not quite enough.

THE BURY INDUSTRY

Just as Mitchells, Ashworth, Stansfield and Company Limited (MASCO) was formed, there was a breakaway movement from Rossendale to Bury, Lancashire, which set up in direct competition by forming the Bury Felt Manufacturing Company Limited. This was led by a Mr Clegg with support from a consortium of Bury businessmen, including J. Hill, the mayor of Bury, and a builder called J. Turner, as well as a number of skilled felt makers from Rossendale. They produced their felt in Hudcar Mill, an old mill built in 1825 that was notable as having once been owned by the textile philanthropist William Gregg of Quarrybank Styal, and they prospered there until 1928 as a sizeable threat to the MASCO combine.

Then, after some financial difficulty, the company embarked on an ambitious programme of expansion, acquiring two nearby mills, one at Bright Street – the old Chesham hat works – and Springfield Mill next to it.

Again the company prospered and by 1962 the Bury Felt Manufacturing Company rivalled MASCO in size and merged with it on 25 May 1962 to

INDUSTRY HISTORY

Scale model of Hudcar Mill, c. 1946.

form Bury and Masco Industries Limited, thereby ending over half a century of rivalry. However, with the continued decline in the industry, the closure of felt mills was inevitable and by 1973 Baltic Mill was the only felt mill left in Rossendale Valley.

Springfield Mill in the foreground with Hudcar Mill in the background, c. 1946.

COOPER AND COMPANY

The company was started in 1896 by Josiah Cooper, became Limited in 1898, and later reformed as Cooper and Co. (Birmingham) Limited under the control of F. G. Bensly. In these early years the company consolidated its position as a specialist in the manufacture of felt polishing wheels (known as 'bobs') as well as supplying other polishing products, eventually expanding and relocating to custom built premises in Brynmawr, South Wales in 1950.

During this time the Bensly family remained in firm control of the company, with direction passing from father to son, Eric Frank Bensly, until 1963 when the family sold it to John Crossley of Carpet Trades, Kidderminster. The company was then sold to Bury and Masco (Holdings) Limited and in 1981 the production was moved from South Wales to Bury, where a new company named Bury Cooper Whitehead was formed.

THE KIDDERMINSTER CONNECTION

There was no history of felt making in the Kidderminster area in Worcestershire until Hebert Smith, known locally as 'Piggy' Smith, decided he would make felt against all advice and against all odds.

He first became General Manager of James Humphries and Sons Limited in 1906, an ailing carpet manufacturer, which he re-established as a profitable business and eventually bought out in 1910. He went on to do the same for Charles Harrison and Sons, this time buying out the Longmeadow Mill facility and converting it from a weaving and needle punching operation into a fully equipped woollen felt operation, growing the business to become 5 per cent of the total UK industry in 1917.

Herbert Smith was a talented man, an accomplished violinist and possessed of considerable energy, which stood him in good stead when he became the chairman of the Carpert Trade Rationing Committee during the First World War, his achievements earning him a baronetcy in 1920. Later when Herbert Smith formed Carpet Trades International the Longmeadow Felt Company became one of the companies in the group, but much later, in 1972, the company was sold to Bury and Masco (Holdings) Limited.

THE PIANO FELT MAKERS

A foremost pioneer of making felt for use in pianos was Richard Jones, who was a felt hat maker based at Crown Mill on the River Wandle in Mitcham near London. He exhibited 'piano felt cloth' and 'canvas for enveloping pianos' at the New York Exhibition of 1853.

In 1859 the company became the Wandle Felt Company, producing surgical as well as piano felts, and in 1871 it was bought out by R. R. Whitehead & Brothers for £20,000. Production remained in Wandle until 1905 when it was transferred to their Royal George Mill at Greenfield, Oldham.

INDUSTRY HISTORY

R. R. Whitehead's Royal George Mill at Greenfield near Oldham.

R. R. Whitehead & Brothers was formed in 1837 by Ralph Radcliffe Whitehead and his three brothers: James Heywood, Francis Frederick, and John Dicken. The natural successor to the business was Francis Frederick's son Ralph Radcliffe junior, but he was disturbed by the conditions at the mill and emigrated to America (where he formed the Byrdcliffe artistic commune of Woodstock fame).

The original company was formed to manufacture and trade in woven woollen cloth and flags, but the company was mainly noted for its Royal George range of piano hammer felts and for its woven felted cloth, which by 1924 was being used extensively for filtration felts in paper making.

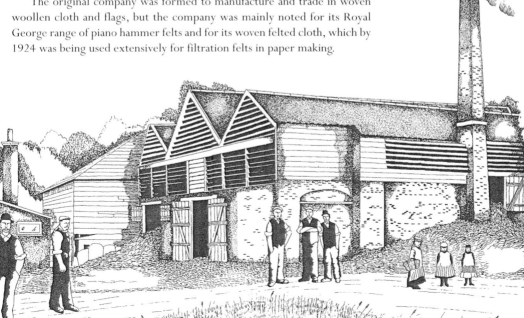

Crown Mill at Mitcham.

Above left:
R. R. Whitehead, from a photograph in the Royal George Mill boardroom.

Above right:
William Naish in 1875.

When the company experienced a sharp downturn in the woven felt market, it was sold to Porritt and Spencer Limited, a company specialising in supplies to paper makers, and which in turn sold out in 1971 to the Scapa Group, whose core business was supplying the paper industry. Eventually R. R. Whitehead lost its independent identity when Scapa integrated it into a new company called Bury Cooper Whitehead Limited. After 1994 it ceased production and from then on Royal George felt was made in Japan.

The only UK company to maintain its independence throughout its history is E. V. Naish Limited, a company important for being one of the earliest producers of piano felts with a worldwide reputation for quality. The Naish family had a long tradition of cloth production going back to the 1800s when William Naish produced corduroy for the farming community at his mill in Crow Lane, Wiltshire. Though he only started felt production in the mid 1850s, by 1869 the company was a world leader in the manufacture of piano hammer felts.

R. R. Whitehead letterhead, c. 1960s.

On William's death his widow Elizabeth Vaudry Naish ran the business and formed the company E. V. Naish Limited, branching out into the manufacture of other felts. The company continued to expand after her death in 1901 and developed production into technical felts and felts for military applications so that by 1960 the company was employing eighty people and by 1979 had 13.5 per cent of the total UK felt production. Unfortunately, by 1994 the company ceased making felt, though it continued its presence in the British felt market by marketing felts manufactured in Europe.

OTHER FELT MAKERS

There were others who tried their hand at manufacturing felt in different parts of the country and they came and went without having a significant impact upon the industry. E. Edmunds of Berryfield, Bradford, tried to go into production through his company, the West of England Woollen Manufacturing Company, in *c.* 1864 but failed due to lack of technical knowledge. Thomas and Mark Hutchinson at Barnes Cray in Kent were a little more successful in making felt carpets, but in 1865 they too were out of business. Hodgetts of Park Side in the village of Cradely, Stourbridge, were quite successful: their business progressed from yarn spinning to hat felt manufacture and then to making polishing felts up to 1876, probably when the company was sold to the Mitchell Brothers of Waterfoot, Rossendale. There was also a small felt company operating in Batley, Yorkshire, in around 1917 called W. G. R. Fox, with a capacity of just one machine.

THE FINAL HISTORY

In 1966 Bury and Masco Limited, the leading UK felt maker, adopted a strategy of acquisition reminiscent of the Mitchell Brothers sixty years earlier, and formed itself into a new holding company incorporated as Bury and Masco (Holdings) Limited. Thereafter this holding company bought out Longmeadow Felt and Cooper and Co. (Birmingham) Ltd from Carpets International and acquired Australian Felts Pty in Australia and Bacon Felts Inc. in America. However, it was in 1978 that the holding company itself was taken over by the Scapa Group, which complemented the felt business that it acquired through its ownership of R. R. Whitehead & Brothers, the piano felt maker. Within two years the Scapa Group amalgamated all its UK woollen felt interests into one new company and formed Bury Cooper Whitehead Limited. This, however, failed to halt the decline in the felt industry and in 1993 the company closed, with much of the capacity going to Germany and the piano felt production and Royal George trademark going to a Japanese company. Sadly the Hudcar site is now divided into industrial units, the picturesque Royal George Mill has been developed as a residential estate, and Baltic Mill has been demolished to accommodate houses.

PROCESSES

INITIALLY the process of making a felt was a relatively straightforward one, consisting of wool blending, hardening, and fulling, but over the years, as felt diversified, there were myriad ways of producing a felt, each giving products of a unique character.

WOOL SELECTION

The selection of wools was probably the most important part of felt production and the particular blend of different wools that might be used to make a particular felt was a closely guarded secret of the wool buyer, who for this reason was often the owner of the company.

The best felting wools were Australian or South African merinos of very fine quality, usually carbonised and bleached, though these could be mixed with a variety of other wools such as those with short fibre lengths, known as noils, to enhance the felting properties. In making the coarse grey felts, exotic wools such as karakuls from as far afield as Pakistan and Afghanistan were used, often supplemented with wools containing short coarse kemp fibres to give the felts more bulk and to stiffen them. Because of the seasonal nature of the supply of wool and the long transit times from the sheep farms, companies had to store large quantities of wool, which led to them building special sizeable stores, which they located well away from production areas as a precaution against fire, as well as for the protection of valuable assets.

The selection of different wool types became ever more sophisticated, as did the techniques of preparing the wool for the main felting processes, giving rise to operations with intriguing names such as teasing, willowing, devilling, scribbling and carding, using such machines as willow, devil, fearnought, scribbler, and card.

Opposite:
The carding shed at Hudcar Mill, The Bury Felt Manufacturing Company, c. 1946.

The process of felt making

Willowing	
Blending	
Scribbling	
Carding	
Hardening:	*Roller hardening • Flat hardening*
Milling:	*Stock milling • Roller milling* *Box milling • Acid milling*
Treatments:	*Proofing • Dyeing • Printing*
Finishing:	*Cropping • Pressing • Sanding*
Fabrication:	*Cutting • Shaping • Laminating*
Despatch	

THE FELT INDUSTRY

Above:
Constituents of a wool blend for a grey felt.

Above right:
A typical felt wool store.

Right:
A devilling machine.

Below:
Albert Mill (left) and the wool store (right), connected by an overhead passage, keeping the wool store separate. The passage was being decorated for the Coronation in 1953.

PROCESSES

PREPARATION

Preparation was necessary because the wool arriving at the mill was matted as well as containing seeds, burr, sand and other debris collected by the sheep in their fleeces in the course of their grazing. All of the machines for preparing the wool were based on the same basic design using a large rotating drum, sometimes called a 'swift', which was covered either with sharp teeth or coarse wires projecting outward.

Surrounding the drum were pairs of spiked rollers with one of them being larger in diameter than the other; the larger was responsible for taking the fibre off the swift, whilst the smaller one put the fibre back on, so that the total effect was one of combing the wool.

Feeding a devil, showing the ducting to the wool bins.

This illustration shows the combing action of worker and stripper rollers around a swift.

THE FELT INDUSTRY

Uniform web coming off the carding engine before being wound on a batt frame.

In the case of willowing and devilling, the objective was also to clear the wool of the debris and make it easier for the scribbling process to concentrate on separating the fibres, whilst the final carding process produced a web of almost totally separated fibres.

CARDING

Although the felt makers used the same carding engines and settings common to the woollen and worsted industry, they had a unique way of collecting the web that emerged from the machine, which symbolised felt production. The web emerging from the card was collected on an cotton belt 40 yards long, which was mounted on a special frame, called a 'batt' frame. This accommodated the belt by running it backwards and forwards over rollers at each end of the frame. At each revolution of the belt an extra layer of fibres

A batt frame at the end of a carding engine; the lap is in the process of being wound up. The shiny black conveyors are to support the fibres when they are upside down.

was added, enabling a sizeable web to be built up as the conveyor continued turning: in practice about fourteen passes were sufficient to make a good felt, after which this fibre batt was wound up into a roll called a 'lap', ready for hardening into a felt.

HARDENING

Historically two forms of hardening developed: roller hardening and flat hardening. Roller hardening was a continuous operation for producing felts up to half an inch in thickness using a machine that was standard throughout the industry, whilst flat hardening was a batch process for producing thick felts (up to 6 inches), which evolved into a variety of different machines.

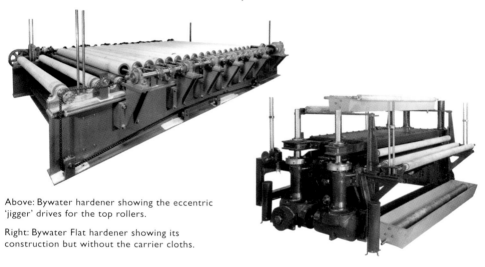

Above: Bywater hardener showing the eccentric 'jigger' drives for the top rollers.

Right: Bywater Flat hardener showing its construction but without the carrier cloths.

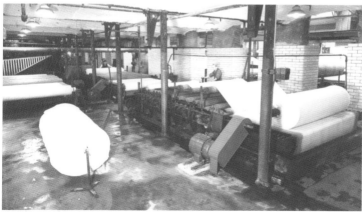

Roller hardener department in Hudcar Mill (c. 1946), showing the cotton conveyor bratt cloths.

THE FELT INDUSTRY

Flat hardener in use at Baltic Mill, c. 1978.

The roller hardening machine consisted of a set of twenty wooden or rubber covered steel rollers on top of a further twenty rollers, some of which were steam heated, and the top rollers were made to jiggle from side to side through eccentric drive shafts. Each set of rollers had a separate piece of heavy-duty cotton, called a 'bratt cloth', surrounding them to provide support for the wool fibres of the lap they passed through the machine and to transmit the vibration into the fibres to felt them.

At the exit end of the machine the two bratt cloths separated and the newly felted fabric had sufficient strength to be peeled off the cloths, rolled up, and handled for the next process.

Below left:
Piano felts drying, showing the taper on the felt sheets.

Below right:
Garside hardeners being used to make Valenki boots at Bury Manufacturing Felt Company, c. 1940.

Flat hardening had a different felting technique and mimicked the action of the early hat felt makers, who felted their wool by laying it on a flat table and rubbing it with a stone. A flat hardener was effectively a massive press with two steam heated platens up to 96 inches wide, the lower platen being fixed whilst the upper platen was capable of being shuffled in a circular motion, known as 'throw', by an amount varying from zero to two inches. The amplitude of the shuffling was carefully controlled to match the thickness of the final felt, the thicker felts needing the greater movement.

For hardening, the fibre batt had to be supported between two cotton or linen carrier cloths, and to make a thick felt, up to ten rolls or laps of fibres were assembled at one end of the machine and fed together between them.

The top platen was lowered with the full weight of several tons resting on the wool and then shuffled for around ten minutes, before being released as a consolidated felt. There were other smaller-scale variants of this machine for carrying out batch processing, such as the yard square table hardener with a tilted top plate to create a felt with a varying thickness and density for making piano hammer felts, or the 20-square-inch Garside hardener.

MILLING

After hardening the felt was milled, or 'fulled', to make the felt denser and stronger using a process similar to that used in consolidation of woven woollen cloth, where wooden hammers were used to beat the cloth. Early machines (called 'stock mills') consisted of two heavy oak hammers hanging like pendulums, with the massive triangular-shaped hammerheads swinging in a cylindrical vessel, shaped to accommodate the roll of felt.

Below left:
A modern paddle stock mill showing the stepped feet, c. 1978.

Below right:
A pilgrim step, roller milling sheet felt.

THE FELT INDUSTRY

A lant cart in Helmshore Mills Textile Museum. Note the old hammer stocks on the left of the picture.

As the industry progressed, many other different types of mill were used, such as positively driven paddle stocks, piston-actioned bumper stocks and box millers. Box millers were used for thin felts, and consisted of a pair of rapidly rotating rollers that continuously flung the felt against a flat plate, the whole machine being enclosed in wooden box casing. For thick felts the felt makers used a specially developed roller miller called a 'Pilgrim Step', where the felt was fed through a set of rollers that oscillated backwards and forwards, squeezing the the felt and inching it slowly through the machine, effectively kneading it in a continuous motion.

They also discovered a method of milling that was capable of making felt as dense as wood by using a three-roll mangle that continuously squeezed an endless band of felt held in a bath of warm sulphuric acid, usually for a considerable period that was measured in days.

The felt makers also fine-tuned the milling process with the use of specialist fulling agents such as urine, for which they had a marked preference, claiming it gave the best felt. They went to extraordinary lengths to collect the urine using a special barrel on wheels, known as a 'lant cart', and visiting every house in the neighbourhood for a contribution; a practice that persisted until the 1930s, when indoor toilets became more widespread and modern soaps, detergents and fulling agents were more effective alternatives.

PRINTING AND DYEING

Once the felt had been made, a multitude of processes were used to tailor the felt to specific end uses, using techniques common to the general textile industry such as printing and dyeing.

For many years block printing was the preferred way of adding design to a felt and even when machine printing was introduced, the former still remained a favoured method of giving design depth, despite the fact that it was a skilled and labour-intensive operation. Many felt makers learnt their trade by starting as tear boys helping the block printer with the menial jobs like mixing the printing pastes or collecting the cow dung; this was used as a base layer in the tear trolley that held the dye paste.

Eventually, however, block printing gave way to continuous machine printing, and when the felt carpet industry disappeared the felt makers gave up printing felt altogether and concentrated instead on just producing dyed felt.

The dyeing process itself was a simple process, requiring just a tub to hold the dye liquor and an overhead roller to keep the felt moving through it, though it was a skilled operation that depended for success on the choice of dyestuffs and the close control of the actual dyeing process in terms of treatment time and temperature cycles. The actual dye recipes used for specific colours were closely guarded secrets kept only by the dyer in a private notebook, not only in the early days when vegetable dyes such as fustic were used, but also when synthetic acid dyes replaced them in 1900. After dyeing, the felts were washed and the surplus water extracted – either by mangling or by spin-drying in a massive hydro-extractor – and then sent for drying.

Block printing Masonic carpet by the last block printer, Herbert Suart, and Ken Ritson (15 years old), the tear boy; note the tear trolley he is preparing for the printer.

DRYING

The early method of drying felts was by hanging and stretching them on tenter frames, which consisted of a fixed upper beam and a moveable lower beam attached to vertical posts or 'posses', each beam being furnished with conically shaped pins raked at an angle in order to pierce and hold the felt in place.

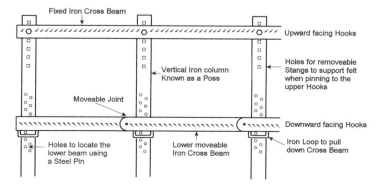

The construction of a hand tenter frame for stretching felt.

The lower beam was jacked downwards with special tools called 'handies' to stretch the felt to a predetermined width, and the beam held in place by pinning the lower beam to the upright posts supporting them.

Initially drying was done out of doors, but later drying was much more efficient when done in specially made heated rooms. This remained the method for very thick and impregnated felts even in modern times when tentering machines, like E. G. Whiteley's, were readily available.

These machine tenters stretched and horizontally dried the felts continuously at high speed, using tenter pins mounted on chains, which pulled out the felt, supporting it as it went through a huge hot air oven.

Above:
Tenter seam at Baltic Mill for impregnated felts, c. 1978, and tenter tools.

Whiteley's Patent Tentering Machine.

FINISHING

The felt makers had a vast range of different wet and dry finishing treatments that they could give to a felt in order to enhance its properties, the most important being moth proofing, rot proofing, and stiffening. The dry treatments were mainly used to alter the surface characteristics using such techniques as cropping or sanding to remove loose outstanding fibres, or to achieve an exact flat surface to meet a precise specification for a technical felt. In the case of thick dense technical felts, huge steam-heated presses were used to ensure that felt thicknesses were achieved to engineering tolerances.

Dense felts could be treated just like wood and could be shaped using wood- and metal-working techniques such as lathe turning, band sawing, sanding, fly pressing, and guillotines to generate felt components of all shapes and descriptions.

For the highest quality felts, particularly superfine felts, steam treatments were used for improving the handle of the felt. These included 'decatising', where the felt was rolled up between a cotton cloth and steam pumped through it, or steam pressing felts continuously on a rotary press to enhance the surface.

Cropping the surface of a felt polishing bob.

PRODUCTS

IN ITS TIME felt was an almost universal material, finding its way into every domestic, textile and engineering market.

CARPETS

Aside from asphalt roofing felt, the mass market for felt was in carpets, rugs and druggets (a form of coarse rug, popular in the Victorian era; initially it was woven wool and linen and was replaced by felt), because in the mid nineteenth century there was nothing to match it for warmth, comfort and appearance at an affordable price, making it readily available to the rapidly growing population. Printing techniques gave them a vibrancy and depth of colour unmatched by other carpets of the time, and at its height in 1863 the felt carpet industry sold over 3.4 million square yards per year, which was 20 per cent of the total carpets produced in Britain. Even when woven carpets eventually took over the market, felt was used to surround them, and when fitted carpets became the norm the feltmakers produced low cost underlay to give woven carpets the same 'spring' as felt.

They also supplied specialist markets such as Masonic carpets, pew seat covers in churches, and as an ultimate accolade, Bury and Masco supplied felt for covering the queen's ballroom, a service for which they were appointed 'Feltmakers to the Queen' with the right to display the Royal Coat of Arms.

A modern offshoot of the carpet production technology was the development of a material for indoor bowling greens, initially for full-length rinks and then as mats for short mat bowling, which was a rapidly expanding market around 1980 when the English Short Mat Bowling Association was set up. This was a fitting epilogue to the felt carpet story.

SLIPPERS

Slippers were a major product success for the felt makers of Rossendale, who converted a simple idea into a multi-million-pound footwear industry.

For some years carders and printers had covered their feet with felt when maintaining the carding engines and printing tables, but it took a housewife,

Opposite:
Fuzzy Felt, showing the box contents.

THE FELT INDUSTRY

Above:
The range of products produced by Mitchells, Ashworth, Stansfield and Company in the early 1900s.

Below left:
Litho print of an advertisement for Gem stair pads, c. 1940.

Below right:
Mitchells, Ashworth, Stansfield and Company's poster for Masonic Carpet designs.

PRODUCTS

Left and below: Felt used for short mat bowling greens c. 1980.

Mrs Wylie, to see the potential of this as slippers and create a cottage industry with the help of Samuel McClerie. However, it took the production expertise of J. W. Rothwell and the marketing skill of Sir Henry Trickett to move it into a national industry with a combination of clever marketing and investment in cutting presses, industrial sewing machines and skilled labour. Once the investment was made, it was a natural evolution from making slippers to manufacturing shoes and as a result Rossendale became the major centre of the shoe industry in Britain, alongside felt making.

An interesting development from the manufacture of slippers was the production of felt boots and although this was not pursued in the United Kingdom it proved very successful for the supply of Valenki boots for the Russian Army; these boots were said to be the saviour of the Battle of Leningrad in the Second World War.

These 'all-in-one' felt boots were ideal for use in ultra low temperature environments where felt outperformed other materials such as leather in terms of insulation and grip on icy surfaces, features readily exploited in climbing boots known as 'Cletterschuh', which had dense felt soles.

THE FELT INDUSTRY

Above left: Advertisement for slipper felt made by the Bury Felt Manufacturing Company.

Above right: Felt slippers, c. 1950.

MILLINERY AND FASHION

There was always a close relationship between the felt and hat industries, but the hats made from sheet felts never quite matched the quality of the hat trade's millinery, where each piece of felt was made individually to a predetermined shape. For felt makers, millinery felt was a steady, if small business that persisted even up to the 1970s, though there was a considerable trade in a low cost felt for the manufacture of novelty seaside 'cowboy' hats, made mostly in Southend.

Along with millinery, the felt makers had a long association with the fashion industry, trying at periodic intervals to break through into the clothing market, but with little mainstream success. In the mid nineteenth century they did have some success in replacing thick woven fabric in pilot jackets, whilst felt clothing is still used in traditional Scandinavian costumes.

The most consistent penetration into the clothing market was through the production of shoulder pads that were made exclusively by the Featherweight Pad Company, which made them using a patented method of production. In the early days of felt making, skirting felts were made and generally worn under other garments for warmth. Apart from this there was only a brief time in the 1950s when circular skirts made of felt became popular, a fashion that came and went rapidly.

Above:
A hat made from printed felt, c. 1950.

Below:
Images from a Bury Felt Manufacturing Company advertisement promoting 'Felt into Fashion', c. 1950s.

Right:
Traditional Scandinavian felt clothes.

PIANO FELTS

A piano mechanism is full of felts of many different types and properties, all making use of felt's excellent sound and vibration control properties: damper felts to absorb vibration, check felts to absorb impacts and most important of all hammer felts to give the optimum sound when striking the piano wire.

Most of these felts could be made in sheets with the minimum of investment and the hardening could be done with small-scale hat felting machinery. This type of manufacture suited the small felt maker, particularly when it came to making hammer felts, since these also required considerable skill to produce and it is to their credit that these specialists could make one piece of felt that had all the different properties required for each note in a piano. This was a feat that the mainstream felt makers could not achieve. Consequently there were only three felt makers noted for hammer felt production: The Wandle Felt Company, R. R. Whitehead & Brothers, and E. V. Naish, though the other main companies did manufacture felt for other parts of the piano movement. Even in modern times, when piano manufacture had passed its peak, the output of hammer felt was considerable, with R. R. Whitehead in 1990 selling sheets capable of making 160,000 pianos.

CRAFTS AND SOFT TOYS

The use of felt for craft and toys was first made popular around 1930 with a series of books by E. Mochrie and I. P. Roseaman, who explored both soft toys and appliqué work. From this a whole new craft industry followed making use of the unique texture of superfine felts and their vast range of colours.

In 1950 Lois Allan launched 'Fuzzy Felt', the creative toy made up of different felt shapes that could be made into pictures by pressing them onto a flocked card. It became so popular that superfine felt thereafter became irrevocably identified as fuzzy felt.

Interior of a piano, showing the hammer felts.

PRODUCTS

Making soft toys from felt maintained its popularity right up to modern times and there were notable children's television characters made from felt, like Barney the dinosaur and Paddington Bear with his colourful felt duffle coat.

EXHIBITIONS

Superfine felts were used most frequently in the exhibition industry, where they were used in large quantities on stands and walls, since it was the cheapest and quickest way of covering surfaces and had the added advantage that when the exhibition was over the boards could be easily restored.

Above:
Bury & Masco exhibition stand promoting craft felt; Bury & Masco colour card.

Left:
Fuzzy Felt, showing the box contents. There was a different theme each year and for 1959 it was the hospital.

47

THE FELT INDUSTRY

Collection of modern felt toys, c. 1978.

However, alternatives such as painting were only marginally more expensive, and there was constant pressure on the felt makers to reduce costs, which they did by diluting the felt with cheaper synthetic fibres such as viscose, until eventually they were replaced altogether by 100 per cent synthetic needlefelts.

MILITARY PRODUCTS

The earliest felt product used by the military was saddle felt, first produced by John Wilkinson in the 1840s to replace woven woollen blankets that the cavalry used under their saddles to emulate the Indian practice of using 'numnahs'. John Wilkinson also established a reputation for gun wads that were used in shells and cartridges to separate the shot from the explosive charge; the felt he manufactured was superior to paperboard in preventing deviation of the bullet as it exited the muzzle.

At the time, these must have generated significant business in supplying materiel for the Crimean War in 1854, the American Civil War in 1861, the Boer wars of 1880 and 1899, and the First World War. By 1980, however, these markets were small: saddle felts were made more for leisure horse riding than for cavalry and modern munitions no longer required felt.

Felt was also used as military packaging for such applications as linings for bomb carriers in order to minimise the risk of explosions, or to line boxes that housed sensitive tools and equipment. Even in 1980 this was a significant business in mostly grey coarse felts, which were specially proofed and produced to a high-level quality specification, usually reserved for aeronautical applications.

TECHNICAL AND ENGINEERING FELTS

At the height of the industry and right up to 1980 felt had found its way into nearly every aspect of engineering; there were few products around the house that did not contain felt, especially if it involved a moving part. It was used as a filter, a wiper, a wick, a seal, a packing material, an insulator, an anti-vibration mounting as well as being used for polishing, lubricating and sound insulation.

There were felt parts in vacuum cleaners and lawn mowers, washing machines, cassette tapes, cars, and even in the high-speed train, whilst when it was used as the nib in pens it gave rise to the concept of 'felt tip' pens.

General industry too, found felt indispensable: metal sheet production relied entirely on a felt wiper for surface protection, the glass industry relied on large circular polishing bobs to give a clear finish, and the automotive industry used it for polishing chrome.

Felt was even used in civil engineering projects, providing anti-vibration mountings for bridge supports and for heavy machinery, which recalled the early use of asphalt-impregnated felts used in the construction of the Menai bridge. These products, branded under the trademarks 'Mascolite' and 'Regalpak', were rot proofed, moth proofed, and impregnated with a special

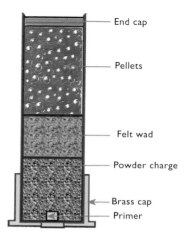

Above:
Diagram showing the construction of a cartridge using a felt wad.

Below:
The different forms of industrial felt.

THE FELT INDUSTRY

Mascolite-impregnated felt for anti-vibration mounting.

Below:
Uses of felt in industry, c. 1970s. Clockwise from top left: Felt washers used in the bearings of the high speed train; a selection of specialised filters, gaskets and polishing wheels; a small square of felt located under the tape in a music cassette; Miniature polishing felts used in dentistry; a felt bearing used in a car air vent; Felt used in a car window channelling.

resin to make them virtually inert. However, they – like the other engineering felt products – have now been largely superseded by rubber or other synthetic materials such as PTFE (polytetrafluoroethylene – the plastic that is used to coat non-stick pans).

LIFE IN THE FELT MILLS

WORKING CONDITIONS in the felt mills were tough at the best of times, since every process involved either steam, water, acid, or other chemicals and at virtually every stage the felts had to be manhandled from one place to another. In the hardening room there was usually so much steam around that it resembled a sauna, and on a cold winter's day the machines generated so much fog that it was impossible to see from one end of the room to the other. The conditions in the dye house were even worse, with the floor permanently wet and swimming with acid that was so corrosive it rotted the dyers' shoes, forcing most dyers to wear clogs. In fact the dye house was probably the most dangerous place to work: incidents of drowning in the dye vats were a fairly common occurrence and there was the ever-present danger of losing fingers and arms when mangling the wet felt.

A felt dyehouse, c. 1940s.

The milling area, though less dangerous, was still an unpleasant, wet place to work, and when urine was used as a fulling agent it must have been obnoxious. However, here the greatest danger was from the unguarded belts that drove the fulling stocks from the overhead line shafting and there were many accidents where workers were caught up in the belting and dragged up to the ceiling.

Even in the drier production areas such as blending and carding, there were significant hazards, mostly from the fibres and dust in the air which were thrown up by the machines, and when unwashed grey wools were used from exotic places like Abyssinia, the risk of infection from anthrax was very real. In the blending and carding rooms there was always the danger of fire because fibres collected wherever there was a flat surface and lint festooned the rafters. Before the mills were lit by electric lights around 1890, the only light was from naked gas lamps and it was common for the flame to surge as it was first lit, often igniting the lint, causing it to flash over and rapidly propagate a fire. Almost every felt manufacturer suffered a mill fire at some time and in the early days they were rarely insured, causing some to go out of business. However, despite all the fires, there were no recorded deaths or injuries to workers.

Perhaps of more concern was the accepted practice up to the 1930s of employing children as young as twelve years old, and there were many instances of them being caught in machinery. Children started work at 6 a.m., worked till 12.30 and then went to school from 1.30 to 5.30, but they also had to work on Saturday mornings, all for a wage of 75p per week in 1919. By comparison, in 1919 the average adult working day was 7.45 a.m. to 5.30 p.m. and Saturday 7.45 a.m. to 12 noon for a wage of £2.75 per week of 55 hours. Although these wages appear low, they were nonetheless in line with those paid by other textile industries of the day.

Apart from the dangers, workers were strictly controlled and monitored to ensure they worked to the maximum, a practice that persisted even up to 1960. Clocking-on procedures and lateness penalties were well established, with each employee having a number identity often associated with a physical brass token that controlled all aspects of their employment.

This strictness even stretched in some mills to the toilet arrangements, there being no doors on the stalls to enable supervisors to monitor staff. Even in 1957 when doors were put on they were of the saloon type so that the occupant could still be seen.

Work token for R. Giddens used by Albert works.

LIFE IN THE FELT MILLS

Albert Works Brass Band with Colonel Mitchell at Ashlands.

In one all-male mill, open urinals were dotted around the mill so that the workers did not need to leave their allotted workplace.

Although these working conditions seem harsh by modern standards, there was a high degree of community spirit in each of the mills fostered by the mill owners who, though strict in their management style, tried their best to care for their employees. Whole families and generations tended to work in the same mill and recruitment was mostly by recommendation of an employee to the mill managers; this personal guarantee ensured a compliant, reliable workforce. During both world wars, employees who joined the services were guaranteed jobs on their return, and during the Depression priority for jobs was always given to those families connected to the mills. Community leisure activities were encouraged and it was common for the owners to take all the mill workers for a day out at the seaside, and for the owners to sponsor activities such as brass bands.

One company even had a nursery to look after the children of their employees. As a mark of respect and the special relationship with the owners the employees always addressed them as 'Mr', to be followed by the first name ('Mr George', or 'Mr Robert').

Despite all the hardship and autocratic management, most employees had an extraordinary loyalty to the company and the owners, and in the whole history of felt making there were no strikes or major disputes, even when mills were under threat of closure, which is a testament to a closely knit community.

FURTHER READING

There are no formal books dedicated to industrial felt making, with most of the information being published by the felt makers themselves as promotional material and therefore not readily available. The following publications have some connection with felt:

Bartlett, Neville. *Carpeting the Millions: The Growth of Britain's Carpet Industry*. John Donald Publishers Ltd., 1977.

Beaumont, Roberts. *Carpets and Rugs*. Scott, Greenwood and Son, 1924.

Burkett, M. E. *The Art of the Felt Maker*. Abbott Hall Art Gallery, 1979.

Cronkshaw, Phyllis, M. A. *An Industrial Romance of the Rossendale Valley: The Development of the Shoe and Slipper Industries*. Manuscript in Rawtenstall Library.

Davies, John. *Mills of Rawtenstall*. Typescript in Rawtenstall Library.

Gordon, Beverley. *Felt Making*. Wilson Gupthill, 1980.

Hawkins, J. H. *History of the Worshipful Company of the Art or Mistery of Feltmakers of London*. Self published 1917.

Kuniczak, Ewa Maria. *Start to Felt*. Search Press, 2008.

Naish, John R. *E.V. Naish: The First 200 Years 1800–2000*. Wilton Graphics, Salisbury.

Newbiggin, Thomas. *History of the Forest of Rossendale*. J. J. Riley, 1893.

Smith, Sheila and Walker, Freda. *Feltmaking, The Whys and Wherefores*. Dalefelt Publications, 2005.

Spencer, Joseph. *The Manufacture of Felt*. Typescript in Rawtenstall Library, 1968.

Trustees of the British Museum. *Frozen Tombs: The Culture and Art of the Ancient Tribes of Siberia*. British Museum Publications Ltd., 1978.

Tupling, George Henry. *The Economic History of Rossendale*. Chetham Society: Remains historical and literary connected with the Palatine Counties of Lancashire and Chester, 1927. New series, Vol. 86.

PLACES TO VISIT

There is virtually nothing left to see of any of the felt mills, nor are there any museums dedicated to preserving the memory of the felt industry. The following have connections with felt and are worth visiting to gain a feeling of what the industry must have been like.

Helmshore Mills Textile Museum, Holcombe Road, Helmshore, Rossendale BB4 4NP.
Telephone: 01706 226459.
Website: www.lancashire.gov.uk/acs/sites/museums/venues/helmshore
This museum has some ancient stock mills, a lant cart, block printing accessories and felt samples.

Rawtenstall Library, Queens Square, Haslingdon Road, Rawtenstall, Lancashire BB4 6QU.
Telephone: 01706 227911.
Website: www.lancashire.gov.uk/libraries/services/local/rawtenstall.asp
The best place to learn about the felt industry; the reference library holds a number of interesting typescripts.

Rossendale Museum, Whitaker Park, Haslingden Road, Rawtenstall, Rossendale BB4 6RE.
Telephone: 01706 260785.
Website: www.lancashire.gov.uk/acs/sites/museums/venues/rossendale
Set in Whitaker Park, this gives a good impression of what a mill-owner's house was like.

Wandle Industrial Museum, The Vestry Hall Annex, London Road, Mitcham CR4 3UD.
Telephone: 020 8648 0127.
Website: www.wandle.org
The only industrial museum in the South East, promoting the history of the industry along the River Wandle.

INDEX

Page numbers in Italics refer to illustrations.

Abott, William 13, *14*
Acid milling 36, *29*
Albert Works 19, *30*, *52*, *53*
Ashworth, Richard 20, 21, *20*
Asphalt felt 7, 13, 41, 49
Australian Felts pty 27
Bacon Felts Inc 27
Baize 9
Baltic Mill 19, 20, 23, 27 *21*
Barcroft, James 17, 20, 21
Batt 32, 35
Batt frame 14, 32
Bensly, Francis, George 24
Blending *29*, 52
Block printing 36, 37, *37*
Bobs, (polishing) 24, 49
Boulinikon Felt Company 21
Bowling green 41, *43*
Box miller *29*, 36
Bratt cloth 34
Burkett, M. E. 5
Bury and Masco 8, 23, 24, 27, 41
Bury Cooper Whitehead 8, 24, 26, 27
Bury Felt Manufacturing Company 7, *7*, 22
Bywater, William 17
Carding *28*, *29*, 32, 52
Carding engine 32, 41
Carpet 5, 7, 15, 16, 17, 19, 20, 24, 27, 41
Carpet, Masonic 37, 41
Carr and Butterworth 16, *17*
Cooper and Co 24, 27
Crown Mill 24, *25*
Decatising 39
Devil 29, *30*, *31*
Devilling 29, *31*
Directional Friction Effect (DFE) 10
Drying, *29*, 37, 38
Dyeing *29*, 36, 37
Elmwood Mill 15
Fabrication 24

Fashion 44, *45*,
Fearnought 29
Felting, by hand 5, 6, 8, *8*
Feltmakers Association 8
Finishing 39
Flat hardener 17, 33, *33*, *34*, 35
Fulling 29, 35, 36,
Fuzzy felt *40*, 46, *47*
Garside Hardener 35
Gun wads 16, 48
Hair felt 9, 13, 16
Handies 38, *38*
Hardener *15*, 17, *33*, *34*, *35*
Hardening 4, *14*, *29*, 33, 34, 35
Hardsough Mill 1
Hat 6, 9, 13, 24, 27, 35, 44, *45*, 46
Hodgetts felt company 7
Hudcar Mill 22, *23*, 27
Industrial felt 49, *49*, 33
Jigging 4
Jones, Richard 4
Lant cart 6, *36*
Lap *32*, 33, *33*, 34, 35
Longholme Mill 20
Longmeadow Mill 24
Lumb Holes Mill 20
Masco 22
Mascolite 50, *50*
McClerie, Samuel 43
McNeill, Forbes 7, 13, 15
Millinery 44
Milling *29*, 35, 36, 52
Mitchell Brothers 17, 18, 19, 20, 21
Mitchells, Ashworth, Stansfield and Company 21, 22
Myrtle Grove Mill 17
Naish, E. V. 26, 27, 46
Naish, William 26, *26*
Needlefelt 9, 17, 48
Needle punching, 19, *19*
Non-woven 7, 9
Patent Woollen Cloth 9
Patent Woollen Cloth Company 15, 16, 21
Piano felt 24, *34*, 27, 46, *46*
Pilgrim step *35*, 36
Plunge Mill 20
Porritt and Spencer 26

Posses 37
Rawlinson, Roland 17
Regalpak 50
Roller hardening 33
Roofing felt 7, *7*, 9, 13, 15, 41
Rostron, Edward 17, *17*, 20
Rothwell, Henry 19
Rothwell, J. W. 20, 23
Royal George Mill 24, *25*
Royal Victoria Carpets 16
R. R. Whitehead 25, 26, 27, 46
Saddle felt 5, 16, 48
Scapa Group 26, 27
Scribbler 29
Scribbling 29, *29*
Shawclough Mill 20
Siss Clough Mill 17, *18*
Slippers 19, 41, *44*
Smith, Herbert 24
Springfield mill 22, *23*
St Clement 6, 7, *7*
St Helens Mill 15, *15*
Stansfield, William 20
Stock mill 35, *35*, 52
Superfine felt 39, 46, 47, *47*
Tear boy 37, *37*
Tear trolley 37
Technical felt 27, 39, 49, *49*
Tenter 37, *37*, 38
Tenter machine 38, *38*
Todd Carr Mill 17, *18*
Trickett, Sir Henry 43
Valenki boot *34*, 43
Wandle Felt Company 24, 46
Waterbarn mill 18
Whitehead, Ralph, Radcliffe 24, *25*
Wilkinson, John 14, 15, 16, 48
Williams, Thomas. Robinson 7, 13, 14, 15
Willowing 29, *29*
Wool 29
Wool, blend 29
Wool, crimp 11, *11*
Wool, fibre 10, 11, *11*, 29
Wool, scales 10, *10*, 11
Worshipful company of Feltmakers 6, *6*, 9
Woven felt 9, 25